COMPETITIVE MATHEMATICS 2

INTRODUCTION

This objective mathematics series provides a basic and challenging problem of mathematics from particular topics. It can be used to brush up ones basics and checking up the preparation level of particular topics. It is equally helpful to the traditional classes as well as competitions. It can be also taken as a revision material for any competition which includes the test of basic mathematics. If you want to grasp the subject before practicing these multiple choice questions, you can go through the website http://www.ncert.nic.in/ncerts/textbook/textbook.htm and down load the free copy of mathematics books and after having command on the topic practice it. For revision purpose, important points are given at the starting of each topic.

CONTENTS

5. SURFACE AREA AND VOLUME

SOME IMPORTANT POINTS

- Surface area of cuboid = 2(lb+bh+hl).
- Lateral surface area of cuboid = 2(l+b).
- Lateral surface area of cube = $4a^2$.
- Surface area of cube = $6a^2$.
- C.S.A of cylinder = $2\pi r h$.
- T.S.A of cylinder = $2\pi r(r + h)$.
- C.S.A of cone = $\pi r l$.
- T.S.A of sphere = $4\pi r^2$.
- T.S.A of hemisphere = $3\pi r^2$.
- C.S.A of hemisphere = $2\pi r^2$
- C.S.A of frustum of cone = $\pi l(r1 + r2)\; where\; l = \sqrt{h^2} + (r1+r2)^2$.
- T.S.A of frustum of cone = $\pi l(r1 + r2) + (r1^2 + r2^2)^2$.
- Volume of cuboid = l*b*h.
- Volume of cube = a^3.
- Volume of cylinder = πr^2h.
- Volume of cone = $1/3\pi r^2$h.
- Volume of sphere = $4/3\pi r^3$.
- Volume of hemisphere = $2/3\pi r^3$.
- Volume of frustum of cone = $1/3\pi h(r1^2+r2^2+r1+r2)$.

5. SURFACE AREA AND VOLUME 1

1. Three cube of each volume 8cm^3 are joined end to end .Find the volume of the cuboid?

    a. 26cm^3　　　　b. 24cm^3　　　　c. 26cm^3　　　　d. 20cm^3

2. Rani made a bird bath for his garden in the shape of a cylinder with a

    Hemispherical depression at one end . The height of the cylinder is 2.46m and Its radius is 60cm .find the surface area of the bird bath?

    a. 115405.72cm^2　b. 32045cm^2　　　c. 112405.79cm^2　d.9234cm^2

3. The length is 125cm and radius 3.5cm .find the area of cylinder?

    a. 1629cm^2　　b. 3729cm^2　　　c. 19250cm^2　　d. 2750cm^2

4. Five cube of each volume 125cm^3 are joined end to end .find the area of the cuboid?

    a. 300cm^2　　b. 400cm^2　　　c. 250cm^2　　　d. 200cm^2

5. Radius of sphere is 6cm .find the volume of sphere?

    a. 805.15cm^3　b. 800cm^3　　　c. 905.15cm^3　　d. 915cm^3

6. A height of cylinder is 9cm and radius is 6cm .find the total surface area of

    cylinder?

    a. 292cm^2　　b. 326cm^2　　　c. 592cm^2　　　d. 565.72cm^2

7. The cylinder and a cone are equal for the diameter and height .find the ratio of

    their volume?

    a. 3:1　　　　　b. 4:2

c. 3:2   d. 2:1

8.   The base and height of a cylinder,a cone and a hemisphere are equal . Find the

     of their volumes?

     a. 2:1:2   b. 3:1:2   c. 3:2:1   d. 2:3:1

9.   A scooty wheel makes 5000 revolutions in moving 10km .find the perimeter

     of wheel?

     a. 5m   b. 10m   c. 9m   d. 2m

10. The volume is 729cm^3 for the cut out of a cube .find the height of largest right

     circular cone?

     a. 27cm   b. 12cm   c. 9cm   d. 3.7cm

11. The area of base is 66cm^2 and volume of a cuboid is 440cm^3 .find the height?

     a. 10/3cm   b. 20/3m   c. 20/3cm   d. 10/3m

12. Volume of the cubes is in the ratio of 8:125.The ratio of their surface area is?

     a. 4:25   b. 6:5   c. 9:2   d. 4:2π

13. Find the slant height of frustum 3cm and 6cm are two base are for the radii and

     height of frustum is 4cm?

     a. 2cm   b. 2m   c. 5cm   d. 9cm

14. The height of cylinder is 7cm and volume of right circular is 448$^\pi$ cm^3 .What is radius?

a. 2cm   b. 4cm   c. 0.4cm   d. 8cm

15.   The frustum of a cone of height 10cm and the diameters of its two circular ends

       are 2cm and 6cm .find the volume of frustum ?

       a. 464.29cm³   b. 544.76cm³   c. 500cm³   d. 4.62m³

16.   Find the area of frustum of cone? The height is 14cm and radius are 3cm and 6cm?

       a. $198\sqrt{277}/7$cm²   b. 89cm²   c. $129\sqrt{26}/7$ d. $129\sqrt{36}/7$

17.   Find the curved surface area of frustum of cone.The radius are 9cm and 12cm and slant height is 15cm?

       a. 870cm²   b. 990cm²   c. 502cm²   d. 860cm²

18.   Find the area of sphere radius is 2m surface?

       a. 50.cm²   b. 47.69cm²   c.35.39cm²   d. 60cm²

19.   Find the total surface area and volume .The length, breadth and altitude of a

       Cuboid 15, 10 and 8cm?

       a. 900cm²,1400cm³   b. 800cm³,1000cm³

       c. 700cm²,1400cm³   d. 700cm²,1200cm³

20.   A cylinder and a cone have same circular base .The heights are equal .find their

       volumes ?

       a.2:1   b. 3:1   c. 4:1   d. 6:1

21.   Find the volume of a sphere and the ratio of the surface area?

a. 2:2r   b. 3:2r   c. 3:r   d. 3:4r

22. The surface area of cube is 225cm^2 .find the one diagonal of the face of a cube?

   a. $12\sqrt{2}$ cm   b. 15cm   c. $15\sqrt{2}$ cm   d. 12cm

23. The sides of a cuboid are doubled then are then its volume will be increased by?

   a. Two times   b. Eight times

   c. Four times   d. Five times

24. Find the volume of a cylinder .The height is 14cm and the diameter of a circular

   base of a right circular cylinder is 40cm?

   a. 17600cm^3 b. 116000cm^3 c. 5289cm^3 d. 6892cm^3

25. Find the surface area of a sphere and the ratio of the volume is?

   a. r:3   b. 2r:3   c. 3:r   d. 3:r

26. The surface area of a sphere is 4^π. Find the radius?

   a. 3cm   b. 2cm   c. 1cm   d. 0.2cm

27. A room of a dimensions is 6m*6m*3m .find the length of longest rod which is

   Placed in this room?

   a. 20m   b. 10m   c. 30m   d. 9m

28. The volume of a sphere is $4/3^{\pi}p^3$.What is diameter?

   a. 3p   b. 2p   c. 4p   d. 6p

29. The solid sphere of radius is 6cm .what is volume?

a. $288\pi cm^2$ b. $360\pi cm^2$ c. $270\pi cm^2$ d. $362\pi cm^2$

30.   The volume of the earth 44m³ and 14m deep .What is diameter?

     a. 3m   b. 1m   c. 2m   d. 4m

31.   The volume of cone is 2/3 $\pi r^2 h$ and radius is r. Find the height?

     a. 3h   b. 2h   c.4h   d.6h

32.   If the radius is R .find the total surface area of hemisphere?

     a. $3\pi R^2$ b. $4\pi R^2$ c. $6\pi R^2$ d. $2\pi R^2$

33.   If the radius and height of a cylinder is 2cm .find the volume of cylinder?

     a. 66cm³   b. 88cm³   c. 77cm³   d.99cm³

34.   The height is 14cm and diameter of base 4cm .find the volume of cylinder?

     a. 374cm³   b. 236cm³   c. 256cm³   d. 176cm³

35.   The cube and the total surface area of a sphere are equal .find the ratio of their

     volumes?

     a. $3:\pi$ b. $\sqrt{6}:\sqrt{\pi}$ c. $2:\pi$ d. $\sqrt{3}:\sqrt{\pi}$

36.   The area of a cone 154 cm .find the diameter?

     a. 8cm   b. 14cm   c. 9cm   d.12cm

37.   Find the volume of cuboid if the side of the cuboid are l, b , h?

     a. l+b+h   b.2 lbh   c.lbh   d.2(l+b+h)

38.   The T.S.A of a hemisphere is $147\pi cm^2$ find the radius?

     a.6cm   b.7cm   c.8cm   d.9cm

39. The volume of a cube is 1728^πcm³. What is the T.S.A.?

   a. 984 cm²   b. 864cm²   c. 295 cm²   d.809 cm²

40. Find the volume of a cone. The height is 28cm and diameter of base is 42cm?

   a. 13742   b. 12936   c.10243 d.12493

41. The volume of a sphere is $4/3\pi x^3$.find the diameter of sphere ?

   a. 4x   b. 2x   c. 3x   d. x

42. The volume of the greatest cube that can be out from the sphere of radius 5cm

   is?

   a. $2000/3\sqrt{3}$cm³   b. $3000/2\sqrt{3}cm3$

   c.$1000/3\sqrt{3}cm3$ d.$4000/2\sqrt{3}cm3$

43. A spherical ball is dipped in water .If its volume is $4/3^\pi$ $(7)^3$cm³ .What is radius?

   a. 5cm   b. 7cm   c. 6cm   d. 4cm

44. The base of a cone and the radii of a sphere are equal .find the height of cone?

   a. 2r   b. 3r   c. 4r   d. 6r

45. How many small cubes of 2cm edge can be made from cube of side 8cm?

   a. 56   b. 24   c. 64   d. 60

46. The length is 10m breadth 8cm and height 6m .find the volume of tank?

   a. 390m³   b. 489m³   c. 480m³   d. 369m³

47.   The length is 12m, breadth 10m and height 9m .find the area of room ?

a. 296m^2 b. 396m^2 c. 260m^2 d.360m^2

48.   The edge of cube is 2a .find the volume of cube?

a. 6a^3 b. 8a^3 c. 4a^3 d. 9a^3

49.   The cube of length of edge is 10cm .find the volume of cube?

a. 100cm^3 b. 1000cm^3 c. 200cm^3 d. 2000cm^3

50.   Find the height of rectangular box .The area is 574m^2 14cm long and
9cm wide?

a. 6cm   b. 7cm   c. 9cm   d. 4cm

51.   The sphere of radius is 6cm. What is surface area of sphere?

a. 260πcm^2 b. 126πcm^2 c. 144πcm^2 d. 269πcm^2

52.   The volume hemisphere of radius is 3cm.find the volume of hemisphere?

a. 18πcm^3 b.12πcm^3 c. 14πcm^3 d. 7πcm^3

53.   The height cuboid is 12cm; length is 4cm, breadth 5cm .find the volume of

cuboid?

a. 360cm^3 b. 260cm^3 c. 290cm^3 d. 240cm^3

Answers:

Q	A	Q	A	Q	A	Q	A	Q	A
1	B	13	C	25	A	37	B	49	B
2	A	14	D	26	C	38	B	50	B
3	C	15	B	27	D	39	B	51	C
4	C	16	A	28	B	40	B	52	A
5	C	17	C	29	A	41	B	53	D
6	D	18	B	30	C	42	C		
7	A	19	D	31	B	43	B		
8	B	20	B	32	A	44	C		
9	D	21	C	33	B	45	C		
10	C	22	C	34	D	46	C		
11	C	23	B	35	B	47	B		
12	A	24	A	36	B	48	B		

6. SURFACE AREA AND VOLUME

1. Total surface area of a cylinder is?

   a. $2\pi r(r+h)$ b. $2\pi rh$ c. $\pi r^2 h$ d. None of these

2. Lateral surface area of a?

   a. $2\pi r(r+h)$ b. $2\pi rh$ c. $\pi r^2 h$ d. None of these

3. Volume of cylinder is?

   a. πrh b. $\pi r^2 h$ c. $2\pi r^2 h$ d. $2\pi r(r+h)$

4. Volume of a sphere is $108/3\pi r^3$, then its radius will be?

   a. 2r b. 3r c. 4r d. 5r

5. The volume of the cube is 75cm^3 . The total surface area of a cube is?

   a. 130 b. 140 c. 150cm^2 d. 160

6. Volume of cuboid is?

   a. l*b*h b. 2h(l+b) c. 2(lb+bh+hl) d. None of these

7. Total surface area of cube is?

   a. 4a^2 b. 5a^2 c. 6a^2 d. a$\sqrt{3}$

8. The diagonal of cube is $6\sqrt{3}$cm its lateral surface area?

   a. 133 b. 144 c. 155 d. 1166

9. Total surface area of cube is 6cm^2, then its volume is?

   a. 1cm^3 b.3cm^3 c. 4cm^2 d. 6cm^3

10. Volume of sphere is?

a. 2/3 πr^3　　　　b. 4/3 πr^3　c. 4 πr^2　　　d. None of these

11. Surface area of sphere is?

a. 2 πr^2　　　　b. 4 πr^2　　c. 2 $\pi r^2 h$　　d. 4/3 πr^3

12. Slant height of cone (l) is?

a. $\sqrt{h^2 + r^2}$　　　　b. $\sqrt{l^2 + r^2}$　c. $\pi r(l + r)$　　　d. None of these

13. Total surface area of cone?

a. $\pi r l$　　　　b. $\pi r^2 l$　　c. $\pi r(l + r)$　　　d. None of these

14. Volume of cone is?

a. 1/3$\pi r^2 h$　　　　b. 2/3$\pi r^2 h$　c. 4/3πr　　　d. None of these

15. Total surface area of cuboid is?

a. 2(lb+bh+hl)　　b. l*b*h　　c. 2h(l+b)　　　d. None of these

16. The volume of the cube is 10648m^3 .The lateral surface of a cube is?

a. 848　　　　b. 858　　c. 484　　　d. 448

17. In a cylinder , if radius and height is doubled , the volume of new cylinder will?

a. Same　　　　b. Doubled　c. Full　　　d. Four times

18. The total surface area of hemisphere is?

a. 2πr^2　　　　b. 3πr^3　　c. 2/3πr^3　　　d. None of these

19. Volume of hemisphere is?

a. 2/3πr^3　　　　b. 4/3πr^3　c. 1/3$\pi r^2 h$　　　d. None of these

20. Lateral surface area of cuboid is?

   a. 2h(l+b)   b. 2l(h+b)   c. 2b(l+h)   d. None of these

21. The surface of the two spheres are in the ratio 27:75. Find the ratio of their surface areas?

   a. 9:3   b. 9:25   c. 3:5   d. 4:5

22. The surface area of a sphere of radius 7cm is seven times the area of the curved surface of a cone of radius 3cm. Find the height?

   a. 4cm   b. 40cm   c. $\sqrt{40}$ d. 0.4cm

23. Find the amount of water displaced by a solid spherical ball of diameter 4.2 cm,

   When it is completely immersed in water?

   a. 15.5cm³   b. 14.8cm³   c. 13.2cm³   d. None of these

24. Curved surface area of hemisphere is?

   a. $2\pi r^2$ b. $4\pi r^2$ c. 2/3 πr^3 d. 4/3 πr^3

25. An edge of a cube measure r cm. If the largest possible right circular cone is cut

   out of this cube, then the volume of the cone is?

   a. 1/12 πr^3 b. 1/6 πr^3 c. 1/3 $\pi r^2 h$ d. 4/3 πr^3

26. Curved surface area of cone is?

   a. 2 $\pi r l$ b. 3 $\pi r l$ c. 4 $\pi r l$ d. $\pi r l$

27. Slant height 2l is $\pi r \left(l + \dfrac{1}{4}\right)$ then, find the total surface area of a cone whose radius

is r?

    a. r/2   b. r/3   c. r/4   d. None of these

28.   The surface area of a sphere of radius 8 cm is eight times the area of the
       curved Surface of a cone of radius 5 cm. Find the height and the volume of
       the cone?

        a. $\sqrt{39}$cm , 163.56cm³   b. $\sqrt{40}$cm , 159.56cm³

        c. $\sqrt{37}$cm , 148.85cm³   d. None of these

29.   Diagonal of cuboid is?

        a. $\sqrt{l^2 + b^2 + h^2}$ b. 2h(l+b)   c. 2(lb+bh+hl)   d. None of these

30.   A sphere and a right circular cylinder of the same radius have equal
       volumes .By what percentage does the diameter of the cylinder exceed its
       height?

        a. 20%   b. 30%   c. 40%   d. 50%

31.   The volume of a cuboid is a diagonal of cuboid is?

        a. $\sqrt{12}$ b. 8   c. $\sqrt{8}$ d. 12

32.   When the radius is doubled surface area of a cylinder increased by____
       times.

        a. 2   b. 4   c. 6   d. 8

33.   When the radius height is doubled the lateral surface area of a cylinder
       increases

        By _____ times.

        a. 2   b. 4   c. 6   d. 8

34.   There are two cuboid boxes having measurements 60*40*50 and
       50*50*50.

Which box requires the lesser amount of material to make?

a. 60*40*50   b. 50*50*50   c. Both same   d. None of these

35.   Surface area of a cube having side 6cm?

a. 6   b. 6^2 c. 6^3 d. None of these

36.   Total surface area of a cuboid includes the area of?

a. 4faces   b. 2faces   c. 6faces   d. 3faces

37.   Ratio between the lateral area and base area of a cuboid is?

a. 1:2   b. 2:1   c. 4:1   d. 1:4

38.   Two cubes each with side b are joined to form a cuboid .What is the surface

Area of this cuboid. What is the surface area of this cuboid?

a. $12b^2$ b. $10b^2$ c. $18b^2$ d. None of these

39.   Base perimeter of a room is 34m and the height of the room is 10m. Find the

area of the four side walls?

a. $34*10m^2$ b. $2*34*10m^2$ c. $34*10*10m^2$ d. None of these

40.   The curved surface area of a cylinder is $100\pi m^2$. Length of the cylinder is 10m.

Find its radius?

a. 10m   b. 5m   c. 20m   d. None of these

41.   Total surface area of cuboid is?

a. 2(lb+bh+hl)   b. 2h(l+b)   c. 2(l*b*h)   d. l*b*h

42. The length of the diagonal of a cube is $6\sqrt{3}cm$, then the length of the edge of

   the cube is?

   a. 3cm   b. 4cm   c. 15cm   d. 16cm

43. If a sphere is inscribed in a cube, then the ratio of the volume of the cube to the

   volume of the sphere will be?

   a. $4 : \pi$ b. $6 : \pi$ c. $4/3 : \pi$ d. None of these

44. A cone, a hemisphere and a cylinder on equal bases and have the same height .

   The ratio of their volume is?

   a. 3 : 2 : 1   b. 1 : 2 : 3   c. 2 : 1 : 3   d. None of these

45. A shopkeeper has one spherical rasgulla of radius 5cm .With the same amount of material, how many rasgulla of radius 2.5cm can be made ?

   a. 5rasgulla   b. 6rasgulla   c. 7rasgulla   d. 8rasgulla

46. A cylindrical roller 2.5m in length, 1.75in radius when rolled on a road was

   found to cover on a road was found to cover the area of 5500m² .How many revolutions did it make?

   a. 50 revolutions   b. 100revolutions

   c. 150 revolutions   d. 200 revolutions

47. A semi – circular sheet of metal of diameter 28cm is bent to form an open

   conical cup . Find the capacity of the cup?

   a. 677.6cm³   b. 677.25cm³   c. 677.4cm³   d. 677.7cm³

48. A right circular cylinder just encloses a sphere of radius r. The surface area of Cylinder is equal to the?

a. Curved surface area of the cylinder

b. Total surface area of the cylinder

c. Curved surface area of the cone

d. None of these

Answers:

Q	A	Q	A	Q	A	Q	A	Q	A
1	A	12	A	23	B	34	C	45	D
2	B	13	C	24	A	35	C	46	D
3	B	14	A	25	A	36	C	47	A
4	B	15	A	26	D	37	C	48	A
5	C	16	C	27	A	38	B		
6	A	17	B	28	A	39	A		
7	C	18	B	29	A	40	B		
8	B	19	A	30	D	41	A		
9	A	20	A	31	A	42	B		
10	B	21	B	32	B	43	B		
11	B	22	C	33	A	44	B		

HERON'S FORMULA

SOME IMPORTANT POINTS

- Area of Δ =1/2*b*h
- Area of Δ by heron's formula
  Semi perimeter (s) = a+b+c/2
  Area of Δ = $\sqrt{s(s-a)(s-b)(s-c)}$
- Area of right angled Δ =1/2*b*h
- Area of an equilateral Δ = $\sqrt{3}/4$ a^2
- Area of an isosceles Δ =b/4 $\sqrt{4a^2-b^2}$ where a is the equal sides and b is the base
- Area of quadrilateral whose sides and one diagonal is given is find by using heron's formula

CONSTRUCTION AND HERON'S FORMULA

1. By using Heron's formula we can't find the area of?

    a. square   b. rectangle   c. circle   d. rhombus

2. The base of a right angled triangle is 4cm and hypotenuse is 5 cm. what is its area in cm^2?

    a. 20   b. 24   c. 6   d. 15

3. The side of an equilateral triangle is 8cm. find its area in cm^2?

    a. $16\sqrt{3}$ b. 32   c. 16   d. 24

4. The area of an isosceles triangle is $8cm^2$. Then its two sides are?

    a. same   b. all are different

    c. all are same   d. none of these

5. The perimeter of an equilateral triangle is 60cm. what is its area in cm^2?

    a. 400   b. $400/\sqrt{3}$ c. $100/\sqrt{3}$ d. $400\sqrt{3}$

6. The length of the diagonal of a rectangle is 10cm. then its other diagonal is?

    a. 8cm   b. $\sqrt{10}$cm   c. 10cm   d. 15cm

7. The unit of area is?

    a. square unit   b. unit   c. cubic unit   d. tetra unit

8. The perimeter of a rhombus is equal to?

    a. $(1/2)\, d_1 * d_2$ b. $2\,(d_1{}^2 + d_2{}^2)^{1/2}$

c. (½) d1 + d_2 d. $2d_1+d_2$

9. If the diagonal of a square is $2\sqrt{2}$cm, what will be its perimeter?

   a. 16cm^2 b. 16cm   c. 4cm   d. 8cm

10. The length of the side of a regular hexagonal is 8cm, what will be its area in cm^2?

    a. 96   b. 48   c. $96\sqrt{3}$ d. $48\sqrt{3}$

11. The dimension of a triangular board is 6cm, 8cm, and 10cm. then the cost of painting at the rate of 9 rupees per cm^2 will be in rupees?

    a. 200   b. 216   c. 248   d. 300

12. The area of the square is equal to the perimeter of the square. Then what is its side is in appropriate unit is?

    a. 2   b. 6   c. 8   d. 4

13. The length and breadth of a rectangle is equal. Then it is a?

    a. rectangle   b. trapezium   c. rhombus   d. square

14. If the area of a square is 81m^2.then length of the diagonal is in meter?

    a. $9\sqrt{2}$ b. 9   c. $9/\sqrt{2}$ d. $18\sqrt{2}$

15. In Heron's formula S =?

    a. a+b+c   b. (a+b+c)^2/2

    c. ((a+b+c)/2)^2 d. (2a+2b+2c)/4

16. If the side of a square is (x+1) unit. Then its area is in unit^2?

    a. ((x-1)^2+4x))   b. (x-1)^2

    c. (x+3)   d. (x^2+1)

17. The side of a rhombus is 8cm. then its perimeter is in cm?

a. 32   b. 36   c. $8\sqrt{3}$ d. $32\sqrt{2}$

18.   The ratio of side of two squares is 5:6. Then the ratio of their area is?

a. $\sqrt{\left(\dfrac{5}{6}\right)}$ cm²   b. 25/36 cm²   c. 36/25 cm²   d. 25/36

ANSWERS:

QUESTION	ANSWER	QUESTION	ANSWER	QUESTION	ANSWER
1	C	7	A	13	D
2	C	8	B	14	A
3	A	9	D	15	D
4	A	10	C	16	A
5	C	11	B	17	A
6	C	12	D	18	B

LINEAR EQUATION IN TWO VARIABLES

SOME IMPORTANT POINTS

- An equation in the form of ax+by+c=0 where a and b both are real number and a and b are not zero
- A linear equation in two variables has infinite solution
- The graph of every linear equation in two variable is a straight line
- An equation of the type y=mx represent a line passing through origin
- The graph of x=a is a line parallel to y axis
- The graph of y=a is a line parallel to x axis
- Every point on the graph of a linear equation in two variable is the solution of the linear equation
- X=0 is the equation of y axis and y=0 is the equation of x axis

LINEAR EQUATION IN TWO VARIABLES

1. Solve the pair of linear equation by substitution method?

   2x+y=14   x-2y=14

   a. X = 5, Y = 2   b. X = 6, Y = 2   c. X = 3, Y = 2   d. X = 5, Y = 5

2. Solve the pair of linear equation by elimination method?

   3x-5y=8   9x-3y=9

   a. X =7/12, Y = -5/4   b. X = 8, Y = 54

   c. X = 8/12, Y = 9/3   d. X = 6, Y = 5

3. Solve the pair of linear equation by cross multiplication method?

   8x+9y= 9   2x+3y= 4

   a. X = 10, Y = 20   b. X = 141/35, Y = 99/55

   c. X = 50/4, Y = 50   d. X = -3/2, Y = 7/3

4. The coach of football team buys 8 football and 9 track suit for Rs 3900. Later he Buys another football and 3 more track suits of the same kind for Rs 1300 . Represent this situation algebraically?

   a. 8x+9y= 3900   , x+3y= 1300   b. 5x+2y= 1800, 5x+8y= 60

   c. 5x+9y= 3900, x+y+3= 1300   d. 5x+9y= 900, 8x+9y= 1800

5. Annu went to a stationary shop and purchase five pencils and two erasers for Rs 9 .Her friend Soni saw the new variety of pencils and erasers with

Annu.She also bought four pencils five erasers of the same kind for Rs 18. Which equation represent it geometrically?

a. 6x+9y=18, 7x+8y= 12   b. None

c. 5x+2y= 9, 4x+5y=18   d. 6x+2y=8,8x+5y=6

6. 7 pencils and 5 pens together cost Rs 50, whereas 5 pencils and 7 pens together Cost Rs 46. Find the cost of one pencil and that of one pen?

a. 3, 5   b. 5, 3   c. 8, 7   d. 6, 7

7. On comparing the ratios a_1/a_2, b_1/b_2 and c_1/c_2. Find out the line representing the pair of equations intersect at a point, are parallel or coincident?

5x-5y+8= 0   7x+5y-9= 0

a. Parallel   b. Coinciedent   c. Intersecting   d. None of these

8. On comparing the ratios a_1/a_2, b_1/b_2 and c_1/c_2. Find out whether the following pair of linear equation are consistent or inconsistent?

3x+9y= 5; 2x+-6y= 7

a. Consistent   b. Inconsistent   c. No solution   d. Both (a) & (b)

9. Half the perimeter of a rectangle garden, whose length is 5m more than it width is 38 m .Find the dimension of garden?

a. 75, 60   b. 21.5,16.5   c. 50,60   d. 98,60

10. Solve 3x+2y= 11 and 4x-2y= -24 and hence find the value of 'a' for which y=ax+3?

a. 33/10   b. -33/10   c. 50   d. 40

11. The difference between two numbers is 26 and one number is three times the other. Find their solutions by subscription method?

a. 13, 39   b. 80,40   c. 39,13   d. 60,50

12. The ratios of incomes of two persons is 7:9 and the ratio expenditures is 3:4 .Ifeach of them manages to save Rs 2000 per month .Find their monthly incomes?

   a. 11000, 10000   b. 18000,14000   c. 2000,1000   d. 16000,12000

13. Use elimination method to find all if linear equations?

   $2x+3y=8; 4x+8y=7$

   a. 5.5,-1   b. 5.5,1   c. -1,5   d. 5.5,11

14. Six years ago, Meena was thrice as old as Sonu . Eleven years later Meena will be twice as old as Sonu. How old are Meena and Sonu?

   a. 8, 10   b. 54, 22   c. 6, 5   d. 22, 54

15. For which value of does the equations given below has unique solutions?

   $9x+9y+13= 0$ $7x+7y+7= 0$

   a. P = 9   b. P = 8   c. $p \neq 9$ d. $P \neq 8$

16. For what value of b will have infinity many solutions?

   $6x+6y- (6-3)= 0$ $5x+Ky-K= 0$

   a. 8   b. 9   c. 10   d. 11

17. For which values of a and b does the following pair of linear equations will have infinite number of solutions?

   $2x+8y= 9$ $(a+b)x+(a+b)y = a+3b-2$

   a. 8,5   b. 0,0   c. 4,5   d. 6,9

18. For which value of k will the following pair of linear equations have no solutions?

   $X+3Y= 1$ $(2K- 1)x+(K+1)y = 2K+1$

a. K = -4   b. K $\neq$ -4 c. K = 8   d. K $\neq$ 8

19. Solve the following pair of linear equation by substitution method

    5x+8y = 9

    2x+3y = 4?

    a. X = 8, Y = 9   b. X=8, Y = 6   c. X = 5, Y = -2   d. X = 5, Y = 6

20. A fraction becomes ¼ when 1 is subtracted from the numerator and it becomes

    1/5. When 8 is added to its denominator .Find the fraction?

    a. 6/8   b. 8/3   c. 9/5   d. ¼

21. Solve the pair of equation 8/x+3/y= 18, 5/x-9/y= -2

    a. 5, 0   b. 0, 0   c. 4, 0   d. 0, 3

22. Solve the following pair of equation by reducing them to a pair of linear

    Equations:

    6/x+2/1/-2= 2, 7/x-1- 5/y-2= 1?

    a. X = 6, Y = 4   b. X = 4, Y = 7.5   c. X = 5, Y   d. X = 8, Y = 10

23. 5 women and 2men can together finish an embroidery work in 4 days while 6

    women and 3men can finish it 3 days .Find the time taken by the women alone

    to finish the work , and also that time taken by 1 men alone?

    a. 2, 3   b. 3, 2   c. 5, 8   d. 6, 7

24. Solve the pair of equation by reducing them to a pair of linear equations?

    1/3x+1/2y=2; 1/2x+1/3y=13/6?

a.1/2,1/3   b. 1/3, 1/2   c. 3, 4   d. 5, 6

25. $a_1/a_2 = b_1/b_2 = c_1/c_2$ is the condition for?

      a. Intersecting lines   b. Parallel lines   c. Coincident lines   d. None

26. $a_1/a_2 \neq b_1/b_2$ is the condition for?

      a. Intersecting lines   b. Unique solution   c. No solution   d. Both (a) & (b)

27. $a1/a2 = b1/b2 \neq c1/c2$ is the condition for?

      a. Parallel line   b. Coincident line   c. Intersecting line   d. None

28.   The graph of two variables of linear equation represent a?

      a. Triangle   b. Point   c. Straight line   d. Curve

29.   If the system of equations 6x-2y= 3 and Kx-5y= 7 has unique solution then K is?

      a. K = 4   b. K = 3   c. K $\neq$ 15   d. K $\neq$ 4

30.   For x = 3 in 2x-3y= 12 the value of x will be?

      a. -1   b. 1   c. -3/2   d. 0

31.   The pair of linear equations is said to be consistent if they have?

      a. No solution   b. Only one solution

    c. Infinitely many solution   d. Both (a) &(c)

32.   On representing x = a and y = b graphically we get?

      a. Parallel lines   b. coincident

    c. Intersecting lines   d. Intersection at (b,a)

33.   For 3x+2y= 4, x can be written in terms of y as?

      a. Y = 4+2x/3   b. Y = 4-3x/2   c. X = 4-3x/2   d. 4-2y/3

34. For what value of p, the pair of linear equation x+2py= 8 and x+y= 6 has a unique solution x = 10, y = -4?

    a. -3   b. 3   c. ¼   d. 6

35. The point of intersection of the lines 2x −y = 6 and x- axis is?

    a. (3, 0)   b. (0, 6)   c. (6, 0)   d. (-3, 0)

36. Graphically y- 2 = represent a line?

    a. Parallel to x −axis at a distance 2 units from x − axis

    b. Parallel to y −axis at a distance 2 units form y − axis

    c. Parallel to x − axis at a distance 2units from it

    d. Parallel to y − axis at a 2units form x − axis

37. If ax+by = c and Kx+ly = n has a unique solution then the relation between the coefficients will be?

    a. am $\neq$ lb   b. al $\neq$ bk   c. al = bk   d. am = lb

38. What is the value of for which (3,b) lies on 3x-2y = 5?

    a. 1   b. 2   c. 3   d. 4

39. Solve 2x+y- = 0; 2x+4y = 16;

    a. 8/3,8/3   b. 8/3 , 16/3   c. 16/3 , 8/3   d. 9/3 , 5/3

40. Solve for x and y 8/x+y + 1/x-y = 2 ; 8/x+y − 5/x-y = -3?

    a. 141/35, 99/35   b. 10, 20   c. 30, 40   d. 50, 60

41. How many variables are there in the equation 8x − 5y +4 = 0?

    a. 1   b. 2   c. 3   d. 4

42. Find the value of y corresponding to x = 5 in the equation 3x + 2y = 9?

a. -1   b. -3   c. 3   d. 5

43. For what value of k the pair of linear equation has no solution 4x + 6k = 9;

    2x + 3y = -11?

    a.2   b.3   c.4   d. 1

44. If C and E denote the temperature of Celsius and Fahrenheit scales respectively, then the following relations holds ; c = 5/9 (E- 44) . Find the value of c when E = 86?

    a. 23.33   b. 25.33   c. 27.33   d. 28.33

45. Find the two digit number if ratio of digits in 3:1 and their sum is 4?

    a. 81   b. 121   c. 31   d. 27/4

46. The triangle is three angles of a triangle is 4:5:9, what are the measures of three angles in degree is respectively?

    a. 40, 50, 90   b. 50, 40, 90   c. 90,40, 50   d. none

47. The sum and difference of ages of Leena and her elder sister is 50 years and 10 years. Find their ages?

    a. 20, 30   b. 30, 20   c. 50, 30   d.60, 30

48. A man has only 50 paisa coins and 25 paisa coin in his purse. If the ratio of coins is 5:4 and amount Rs 14 in all, how many coins of each type does he have?

    a. 20,16   b. 25,20   c. 10,8   d. None of these

49. Sum of two number is 53 and their difference is 31. Find the two numbers?

    a. 42, 11   b. 10, 20   c. 50, 70   d. 60, 50

50. A father is 6 times as old his son. After 20 years his age will be twice as of his son. Find their presents age?

a. 30, 5   b. 24,4   c. 48,8   d. None of these

51. A number is much greater than as it is less than 25. Find the number?

    a. 21   b. 20   c. 22   d. 18

Answers :

Q	A	Q	A	Q	A	Q	A	Q	A
1	B	12	B	23	D	34	C	45	C
2	A	13	A	24	B	35	A	46	A
3	D	14	B	25	C	36	C	47	A
4	A	15	C	26	D	37	B	48	
5	C	16	B	27	A	38	B	49	A
6	B	17	B	28	C	39	A	50	
7	A	18	B	29	C	40	A	51	B
8	D	19	C	30	C	41	B		
9	B	20	D	31	D	42	B		
10	B	21	B	32	C	43	D		
11	C	22	B	33	D	44	A		

ARITHMETIC PROGRESSION

SOME IMPORTANT POINTS

➤ The general for of an A.P is a, a + d, a +2d, a + 3d---------
➤ Common difference (d) =a^2-a , a^3-a_2, a_n+1-a_n--------------------------
➤ In and A.P, if 1^{st}term is (a) and common difference (d) the then the nth term is given by (a_n = a+$_{(n-1) d}$)
➤ The sum of nth term of an A.P is given by----------------

$$S n = n/2 [2a+ (n-1) d]$$
$$S n= n/2[a + a (n-1) d]$$
$$S n = n/2[a + an]$$
$$S n =n/2[a + l] \qquad \text{(where is the last term)}$$

➤ If a, b and c are in the A.P then ;
2b=a + c
➤ If three number are is an A.P the is given by-----------
a- d, a, a + d

ARTHIMETIC PROGRESSION

1. For the A.P, -8, -4, 0, 4, 8........... write the first term and common difference.

   (a) 8, -4

   (b) -8, -4

   (c) -8, 4

   (d) -8, 2

2. -5, -3, -1 ....it is an A.P then write the next two numbers.

   (a) -3, -5

   (b) 3, 5

   (c) 4, 5

   (d) 2, 3

3. Write first three terms of an A.P whose first terms is 8 and common difference is -2.

   (a) 8, 6, 4

   (b) 8, 10, 12

   (c) 10, 12, 8

   (d) 8, 12, 16

4. write the common difference and first term of an A.P 3, $3+\sqrt{3}$, $3+2\sqrt{3}$, $3+3\sqrt{3}$..............

   (a) 3, $3+\sqrt{3}$

   (b) 3, $\sqrt{3}$

   (c) 4, $4+\sqrt{4}$

   (d)3, $5+\sqrt{3}$

5. (a) find 10th term an A.P 2, 9, 16................

   (a) 60

   (b) 63

(c) 65   (d) 67

6. Which term of an A.P 6, 8, 10...............is 48.

    (a) 20th (b) 30th

    (c) 22th (d) 21th

7. Determine the A.P whose 3rd term is 7 and 7th term is 5.

    (a) 8, 15/2, 14.............   (b) 8, 10, 12.........

    (c) 8, -15/2, -14...........   (d) 8, 9, 10...........

8. How many two digit numbers are divisible by 5.

    (a) 19   (b) 18

    (c) 20   (d) 17

9. How many three digit number can be divisible by 6.

    (a) 100   (b) 50

    (c) 150   (d) 90

10. Find the 10th term from the last term of the A.P:-

    10, 7, 4, ...-62

    (a) -32   (b) -35

    (c) -38   (d) -41

11. In a room there are 24 rose plant in the first row and 22 in the second and 6 in last row the then find how many roués in room.

    (a) 10   (b) 12

    (c) 14   (d) 16

12. Find the missing number of an A.P

    3, ■, 27

    (a) 27   (b) 27/2

    (c) 28   (d) 25

13. Which team of A.P 3, 8, 13, 18..............is 83.

    (a) 16   (b) 18

    (c) 17   (d) 19

14. Find the 30th term of an A.P whose 11th term is 73.

    (a) 173   (b) 175

    (c) 170   (d) 171

15. Find the sum of firs 25 terms of A.P 8, 3, -2........

    (a) 100   (b) -1100

    (c) -1300   (d) 1000

16. The sum of three number of an A.P is 24. The middle term is ....

    (a) 6   (b) 8

    (c) 3   (d) 2

17. If nth term is 3n+7, then what is the of 11th term of A.P is

    (a) 30   (b) 40

    (c) 50   (d) 60

18. If nth term of the A.P 3, 6, 9, ...is 81, then value of n is

    (a) 29   (b) 28

(c) 27   (d) 26

19.   10th term of the A.P x-7, x-2, x+3 is

       (a) x+62   (b) x-48

       (c) X+48   (d) 10x-38

20.   Common difference of A.P is 8 1/8, 8 2/8, 8 3/8...........is

       (a) 1/8   (b) 1 1/8

       (c) 8 1/8   (d) 1

21.   6 2k-3, 10 are in A.P then the value of k is

       (a) 11/3   (b) 11/2

       (c) 11/4   (d) 11/5

22.   Radha saves RS.3 on day, RS .5 on day 2, RS 7 on day 3 and so on. How much money she saves in the month of febuary in leap year.

       (a) 602   (b) 532

       (c) 508   (d) 899

23.   The sum of first six terms of an A.P is 42. The ratioof the 10th term to the 30th terms in 1:3. Find term and 11th term of the A.P.

       (a) , 11th term =22   (b) 2, 11th term =22

       (c) 8, 11th term =10   (d) 6, 9th =8

24.   Find the sum of the add numbers between o and 50.

       (a) 625   (b) 725

       (c) 925   (d) 825

25. Find the first three terms of an A.P whose an is defined as

    $a_n = 3+5n$

    (a) 8, 12, 16   (b) 8, 13, 18

    (c) 8, 13, 15   (d) 8, 14, 20

26. In an A.P a=8, d=3, an=8. Find n.

    (a) 15   (b) 16

    (c) 14   (d) 13

27. Find the sum of first 100 positive integers.

    (a) 5108   (b) 5050

    (c) 7106   (d) 8324

28. The first term of an A.P is 10 and the last term is 20 and the total number of team is 21. Then find their sum.

    (a) 315   (b) 700

    (c) 500   (d) 600

29. The sum of the 6th and 10th term is 44 and the sum of 4th and 8th term is 24. Find the first three terms.

    (a) 38   (b) 39

    (c) 37   (d) 40

30. How many multiple of 8 lies between 10 and 250

    (a) 40   (b) 30

(c) 20 (d) 50

31. The sum of firs 24 terms is 996 and the sum of first 8 term of an A.P is 140. Find the A.P

(a) 5, 8, 7………… (b) 7, 10, 13,16…………

(c) 8, 10, 12………… (d) 9, 8, 7, 6………

32. Find the sum of all the three digits number each of which remainder 3 when divided by 5.

(a) 890990 (b) 75990

(c) 99090 (d) 87650

33. The sum of n terms of two A.P is are in the ration 3, 3n+8:

7n+15. Find the ratio of their 10th term.

(a) 38:85 (b) 39:75

(c) 40:95 (d) 69:5

34. What is the relation between 37th and 17th term of an A.P is zero

(a) Twice (b) ones

(c) Thrice (d) five times

35. What is the value sn-sn-1 if sn and sn-1 are given.

(a) T_n (b) $T_n - T_{n-1}$

(b) $T_n + T_{n+1}$ (c) $T_n - T_n$

ANSWERS:

Q.	A.	Q.	A.	Q.	A.	Q.	A.	Q.	A.
1	C	8	B	15	C	22	D	29	C
2	B	9	C	16	B	23	B	30	B
3	A	10	B	17	B	24	A	31	B
4	B	11	C	18	C	25	B	32	C
5	C	12	B	19	D	26	A	33	A
6	C	13	C	20	A	27	B	34	C
7	A	14	B	21	B	28	A	35	A

NOTES

www.ingramcontent.com/pod-product-compliance
Lightning Source LLC
Chambersburg PA
CBHW080619180526
45168CB00007B/2977